The Incompleat Sound Operator

A Brief Compendium of Recommendations, Tips and Techniques for Sound System Operators at Live Music Performances that use Sound Reinforcement

Why You Should Read This Book

"Doing sound" for a dance, coffeehouse or any other musical or stage event is a complicated job. It requires technical expertise, a bit of muscle, diplomatic skills and a good ear for the kind of sound required for the dance, performance or event.

As a theatre technical director, coffeehouse volunteer, dance sound engineer and sound coordinator, I've been "doing sound" for decades. Along the way, I've learned a few things that I think are worth sharing.

This little book provides one fairly experienced sound person's suggestions, recommendations, hints, tips and rules for doing a good job when you're the person "doing sound."

My focus is on one big, overall goal: Everyone in the house is happy, satisfied, relaxed and having a good time without paying the slightest bit of attention to what you, the sound person, are doing. Everything about the setup, the arrival of the performers, testing, sound check and performance goes so smoothly and effortlessly that nobody is paying any attention at all to the sound operator.

Achieving that goal takes preparation, good technical knowledge, efficient work procedures, troubleshooting ability, a calm demeanor, and good people skills. And of course, you need to have a good ear for, and appreciation of that particular type music or performance, and be able to use the tools at your disposal to help everyone hear it without calling any attention to how you're doing your job.

If you are interested in a reasonably non-technical guide filled with tips and recommendations for how to do a great job when you're "doing sound," this book should be of interest to you.

The Incompleat Sound Operator

A Brief Compendium of Recommendations, Tips and Techniques for Sound System Operators at Live Music Performances that use Sound Reinforcement

by Ridge Kennedy

Hedgehog House
West Orange, New Jersey

Hedgehog House

28 Yale Terrace
West Orange, New Jersey 07052

973-400-9738 • www.hedgehoghousebooks.com

ISBN

978-1-951989-00-2

Publisher's Cataloging-in-Publication Data

Library of Congress Control Number Information
LCCN:201992039

Typeset by Affinity Publisher

Copyright © 2020 by S. Ridgway Kennedy

All rights reserved. No part of this publication may be reproduced, distributed, or transmitted in any form or by any means, including photocopying, recording, or other electronic or mechanical methods, without the prior written permission of the publisher, except in the case of brief quotations embodied in critical reviews and certain other noncommercial uses permitted by copyright law. For permission requests, write to the publisher, addressed "Attention: Permissions Coordinator," at the address above.

10   9   8   7   6   5   4   3   2   1

Contents

Preface 1

Introduction 3

Advance Preparation 5

The Signal Chain 11

Setting Up 13

Microphones and 'Directs' 21

Working with Artists 29

Gain Structure 31

Sound Check 37

Sound Equalization (EQ) 41

Running the Show 49

Big Band Sound 51

Putting Your Gear Away 55

Visual Glossary 57

Resources 63

Sound Checklist 64

Feedback (the Good Kind) 67

Acknowledgments

Here are my thanks: To the mentors, guides, co-conspirators and enablers who have guided one Incompleat Sound Operator's journey. The voyage began in the Department of Speech and Theatre at Hiram College where I learned that you could earn college credit for hanging by one hand, 30 feet above the auditorium floor, to adjust the top shutter on an 8-inch Leko (an ellipsoidal reflector spotlight with an 8-inch diameter lens to the uninitiated). The sound system in the theatre was primitive by modern standards but it was good enough to help me learn many of the basics.

The Folk Project and the Folk Project Folk. Doing sound for dozens of coffeehouse performances provided in-depth experience working with acoustic music and musicians, proceeding systematically and above all, keeping the show running on time.

The Princeton Country Dancers and sound guru Paul Prestopino. Paul specified the gear and I learned to work with it. More importantly though, I had the opportunity to work alongside Paul on many occasions. Paul's vast experience as both a recording and live sound professional as well as his perspective as a brilliant musician have, I believe, helped shape my approach toward "doing sound." In addition to his audio tech and musical expertise, Paul is also one of the kindest and most generous people I have ever known. Having the opportunity to be roadie for Paul has been one of the great privileges of my life.

My thanks to the Thursday Night Dance community of Philadelphia. This is where I graduated to sixteen channels, dove deeper into sound tech and discovered the joy of mixing for really big pick-up bands. I took a hand-off from Bill Quern, a great sound operator and even greater musician who would rather be playing than mixing, to become the sound coordinator for the group. It's my good fortune to be now surrounded by a large group of excellent sound operators, some old timers and many new to our group. Together we provide a level of support to dance bands and dance callers that meets audio pro standards. As the technology demands of dance bands have grown, we have learned and grown together. Thank you Anne, Bill, Bob P, Brian, Jan, Jeffrey, Jim, Jocelyn, Phil, Rob, Scott, and Trisha. It's been a great experience.

Bob Henshaw coordinated sound volunteers for The Flurry in Saratoga Springs for many years and included me in his cadre. I got to do some mixing and stage managing which made it an enjoyable weekend for me. Josh Snitkoff received good enough reports on my work to promote me to the pro sound staff as a relief engineer, the apex of my incompleat sound career. Thank you, Josh and Flurry.

Special thanks to early readers of my manuscript including Susie Lorand, Scott Higgs, Jamie Platt, Marty Brenneis, Joe Cline and Paul Prestopino. They provided valuable comments, corrections and improvements. After picking up the project again, Ellen Wolff, Weogo Reed, Jamey Hutchinson and Scott Higgs provided additional input.

Finally, an acknowledgment to all the sound operators I have met everywhere; at theatres, concerts, coffeehouses, camps, dance venues and online. I'm confident many of you will disagree with some of the things I've said here, or you'll have additional insight, helpful tips and suggestions for better ways to do things. I acknowledge that my education is incompleat and I hope you will share your thoughts and experience with me as I've requested in the final 'Feedback' chapter of this volume.

Now, the sound check is complete. Let's start the show.

Preface

My first gig as a sound operator came in 1967 as the stage hand on duty at the time in Hayden Auditorium at Hiram College. My job: Set up lighting and sound—one speaking microphone—for a world-renowned classical pianist, Rosalyn Tureck.

She let me know, in no uncertain terms, that my technical support for her performance was a disaster. The lighting was poor, the Green Room unsatisfactory and even the single microphone she used for speaking with the audience was inadequate. I got assistance from other people on the theatre staff and I was relieved to know that it wasn't just me—none of us could make things right for Ms. Tureck. The concert and the interactions with the artist lived large in our memories for the rest of the year.

We remember the disasters. We forget our good work; just doing our job, y'know. And the audiences and participants in live sound events recall nothing of the best work we do. It calls no notice to itself. That's what we're after: A perfect, seamless performance where sound reinforcement, just as much as is needed, enhances the performance and the performer's experience without calling attention to itself.

I have been a sound designer and operator for theatre productions. I've set up and operated sound systems for hundreds of live music coffeehouse concerts and participatory dances. My background includes specifying and installing sound gear for theatres, classrooms and dance venues. I have a smattering of experience working on sound recordings and a bit more experience using recorded sound for dance programs.

About fifteen years ago, based on my experience at the time, I wrote a manuscript that tried to consolidate all the basic information I had acquired over the years. The idea was to keep it simple and share information that could be used by anyone running live sound events no matter what specific type of equipment they were using. Then my manuscript went into a "to do" stack. It resurfaced a few months ago, and after reviewing it again, I made a few additions and expanded my horizons beyond dance music to include coffeehouses and other live, primarily acoustic performances and events. I have sought a few more reviewers and present it to you for your further consideration.

Ridge Kennedy
West Orange, New Jersey
April 2020

Introduction

This publication has been created, in particular, to assist new sound operators with the setup and operation of sound systems used at live music events where sound reinforcement is used. It is also my hope that experienced sound engineers, even sound pros, will take the time to read it, consider my perspective and share their thoughts with me as I suggest in my solicitation for 'Feedback' on the last page of the book.

This book doesn't go into detail about specific equipment. It assumes that you, your group or the venue you are using has a sound system. And if the gear is new to you, there are people available to show you how the system is typically set up. It imagines that you have some cabinets or carts or suitcases or plastic milk crates filled with sound gear and that, when you plug everything together correctly, it works.

My goal here is to share a set of tips and techniques that will help you set up and use that equipment efficiently and effectively. The "bigger picture" goal is to suggest to you that it's not all that complicated, but there are a lot of moving parts and some of them don't have anything to do with what's coming out of the speakers. I'd like to assist you in being a very good sound operator. My sincere hope is that this small volume will boost your confidence, impart useful information, encourage your participation and involvement with your group, and help make "doing sound" an enjoyable, rewarding part of your life.

It is true that you're dealing with technology that can be quite complex with subtle artistic nuances that can be compared and discussed ad infinitum. But even if you're a newcomer to sound technology, you can focus on the basics, keep it simple and do a great job.

Here's how.

4 The Incompleat Sound Operator

Advance Preparation

When you accept responsibility for setting up and operating sound for a dance or performance, you know you're going to make an investment in time and energy. So, you have a choice: Plan ahead to gain the confidence of the performers and be well prepared to handle any special requirements; or leave things to chance and hope for the best.

◄ **Talk to the band/performers**

If you know you're going to be "doing sound" for an upcoming event, you can make your job much easier by doing one simple thing: Talk to the musicians/performers. Communicate by e-mail or call the bandleader up. Ask for the sound requirements. The musicians or performers will be delighted to help you.

◄ **Have a sound system inventory/description**

When you communicate with the musicians or other performers, it's a good idea to have a detailed list of the sound equipment you use and a statement about what kind of audio reinforcement support they can expect. A major goal for both the sound technician and the musician: No surprises.

Your inventory/description should include information about:

- The sound board: How many channels and what kind of inputs? XLR, quarter-inch, other?
- Monitors: The speakers that face the musicians and allow them to hear themselves better. Do you have monitors? What kind? How many?
- Monitor mixes: Do you just have one mix, or can you provide more than one if musicians request them?
- Microphones: How many of what type?
- Cables: How many/what kind? XLR? Quarter-inch instrument cables? DI boxes?

If any of the above terminology is unfamiliar, you may want to review the information in the Visual Glossary starting on page 57. Many of the musicians you work with will be savvy sound tech experts themselves. If you've got a good description of your system, you can send it to them so they'll know exactly what to expect at your venue. The purpose of this is to be sure that neither you nor the musicians have an unpleasant surprise when the time for the event arrives. If you can share information about the equipment in advance, everyone has a chance to adjust plans or make arrangements, so you won't have last minute problems.

◀ **Get a picture of how the band sets up**

I like to ask the musicians I communicate with to give me a "lineup." I ask them to list, from left to right as seen from the dance floor or audience, the order of the musicians and how many microphones/microphone lines will be needed. Will musicians need DI (direct input) boxes? Will they be using effects boxes? Keyboards? And what kind of stage monitors do they expect?

Get the best picture you can of what the stage is going to look like. Maybe one of the band members can even send you a "stage plot"—a diagram that indicates where the musicians expect microphones and monitors will be on the stage. When you arrive the day of the event, you'll be able to get everything roughly in place even if the performers have not yet arrived.

◀ **Do everything you can to avoid a last-minute rush**

In my twenties, I did sound for theatre productions and stage events. Years later, I got involved in doing sound for dances because, as a dancer and as a caller, I became frustrated when dances started late. When there were three- or four-person bands, the musicians would be on hand and dancers were in the house, but the last-minute details of setting up sound kept delaying the start of the dance. On nights when there were open bands with dozens of musicians and every microphone channel in use, the delays were worse.

Everyone was hurried. The dancers were waiting. The musicians weren't having fun. I felt there was a need and got involved as a sound operator.

Even now, when I run into problems—feedback in the monitors, a monitor speaker not working or some other glitch—the most frequent root cause of the problem is rushing; trying to do too many things too quickly.

Advance Preparation 7

Everything you do to eliminate rushing will make your job easier. When glitches arise, as inevitably they will, you'll have more time to deal with them. And there's another big benefit. Everyone, and especially the musicians, will be more relaxed. Be on hand with your equipment set up, chairs in place and ready to go when they arrive at the venue. Musicians and other performers will appreciate your attention, feel respected and together, you'll set a positive tone for the entire evening.

◄ **Got tools?**

If you start working regularly as a sound person, you may find times when you need a screwdriver or pliers or a bit of tape. If your group doesn't have a tool box with essential tools in it along with a few spare batteries of various important sizes, consider getting one for the group or putting together your own. On the night that a member of the band says, "Gee, does anybody have a mini-Phillips head screwdriver?" you'll be a hero.

Following are recommendations for a couple of specialized sound tools and and suggestions for the contents you might want to include in a sound tool kit.

◄ **Cable tester**

A cable tester is one specialized tool you might want to get for your group. It has connectors that allow you to plug in both ends of microphone cables, speaker cables and other types of audio cables. It will tell you if the cable is good and how it is wired. You can wiggle the connectors using the tester when you are trying to find the source of an intermittent fault. Good units are available for between $20 and $40.

◄ **dB meter**

A dB (decibel) meter measures sound loudness. It can be a handy tool to use to keep volume levels consistent. Personally, I run sound fairly hot. So, over the years, I've used a dB meter to check the volume and make sure it's not *too* loud. I have an ancient RadioShack® digital meter that has a function which allows it to listen over a user-selectable time and report the high and low readings. Today, you can get inexpensive meters with min and max functions that constantly update the high or low reading. You can get free dB meter apps for smart phones, too. None of these inexpensive tools will be precisely accurate, but they will give a good way to know how your volume levels compare from one event to the next.

Tool Kit Suggestions

Here are a few tools that I have in my tool kit—a small, canvas fishing tackle bag—that that have proven useful to me.

- Channellock Pliers: Adjustable, slip-joint pliers that give you lots of leverage for working, in particular, with microphone stands. If you want to be patriotic, buy the real deal: Channellock® brand pliers, fiercely made in Meadville, PA.

- A pair of all-purpose pliers, a pair of needle nose pliers, and a pair of side cutters.

- Adjustable Wrench: An 8-inch Crescent® adjustable wrench will meet occasional needs for handling nuts and bolts.

- Screwdrivers: One medium and one small in both Philips head and straight blade varieties. An additional extra-small Phillips head. Plus, a set of micro-sized screwdrivers. Look at the screws used on microphone connectors and you'll see how small they need to be.

- Utility knife: A.k.a mat knife. With a sharp blade, and extra blades stored inside the handle.

- Electrical tape: At least one roll of black tape and maybe a few rolls of colored tape if you want to code your cables.

- Flashlight (a torch, if you're in the UK): Mobile phones will work in a pinch but a good flashlight is an essential.

- Batteries: AA and 9V for sure; and possibly some AAAs.

- Adapters: I have two that have come in handy: An XLR female to quarter-inch male and a quarter-inch female to XLR male.

- External Sound Source Adapters: This will vary based on your board. You want the cables and adapters needed to take a signal from the eighth-inch stereo jack found on phones, for example, to connect to your board. Some boards have RCA/phono inputs for "tape in." Or you may need to connect to an unused channel. Figure out how to play music from a mobile phone with an eighth-inch inch jack through your board. You'll have the gear you need.

10 The Incompleat Sound Operator

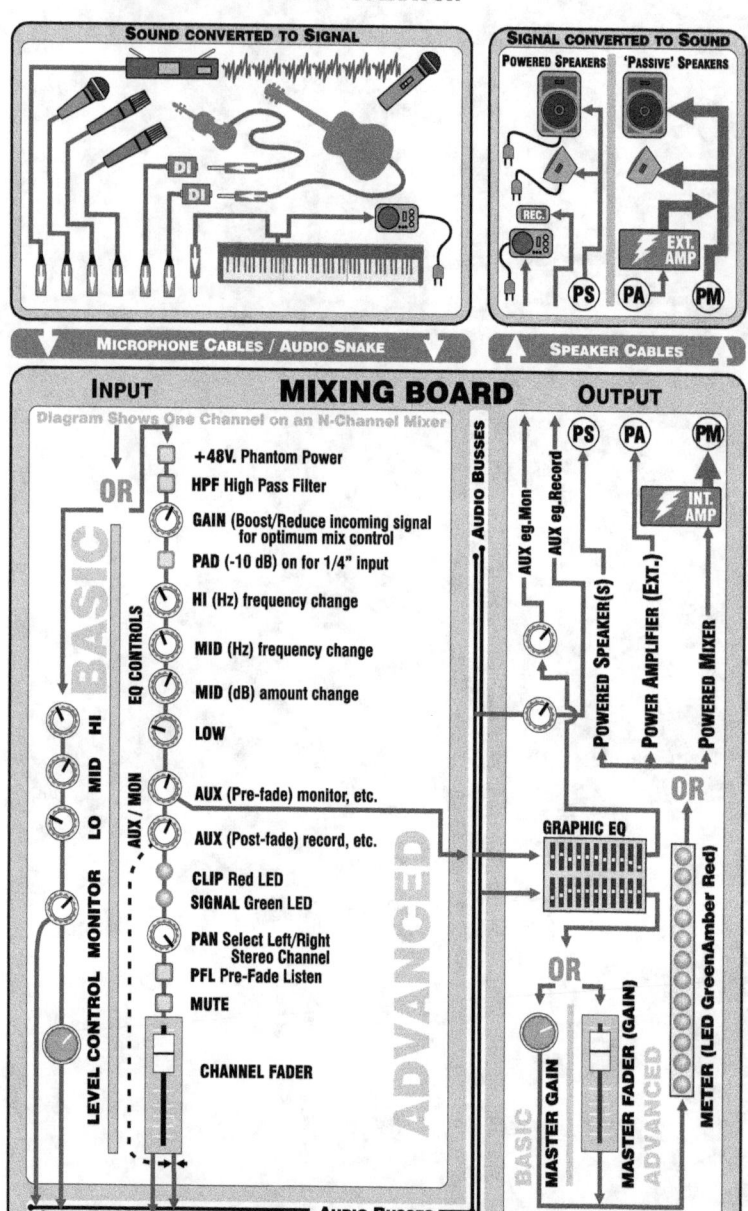

Note: The top item in the top right part of the diagram represents a wireless microphone connection. A transmitter in the body of the handheld microphone body converts sound to radio waves. A receiver converts the RF signal back into an audio signal and sends it to the mixing board.

The Signal Chain

The diagram on the facing page is designed to give new sound operators a graphic reference for the "signal chain"—the conversion of sound into electric current, its passage through the sound reinforcement system and conversion of electricity back into sound. It may be a helpful visual reference for much of the material that follows.

Upper Left: Sound waves are converted into electrical signals by microphones or a contact pickup as with an electric guitar. Or an electronic instrument like an electric piano creates an electrical signal corresponding to sound. The top figure in the diagram shows a wireless microphone using radio waves instead of wires for part of the signal chain. These low-level electrical signals are communicated to a mixing board via microphone cables, or in some cases, an audio snake.

Lower Left: The audio signal goes to pre-amplifier, commonly built in to a mixer. Each signal is controlled by an individual "channel" in the mixer.

Bottom Left and Center: Audio busses (wired or digital connections) in the mixer allow the signals from many channels to be combined into one or more audio "mixes."

Lower Right: The mixes are processed by additional internal audio components. Their power levels are adjusted by "master" controls and the signals are then sent to power amplifiers that are either: 1) built into the mixer (PM in diagram), 2) separate system components typically located adjacent to the mixer (PA) or 3) built into the speakers (PS).

Upper Right: The amplifiers increase the strength of the mix signals to a power level sufficient to "drive" speakers. 'Passive' speakers (right side, upper right) are connected to amplifiers via speaker cables. The low-power mix signals are delivered to powered speakers (left side, upper right) via speaker cables or microphone cables.

The electrical signals move the diaphragms of the speakers, which move air to create audible sound waves.

12 The Incompleat Sound Operator

Setting Up

Assembling Your Equipment and Setting the Stage

Arrive early and develop a routine for setting up the equipment. And for extra emphasis, let me repeat: *Arrive early.*

- **Start with the big stuff**

Put the big things where they're going to go. Put the speaker stands and main speakers in place. Put monitors where they'll be. Put the sound board in its usual spot. If you have a keyboard musician in the band, move the piano or set up the keyboard in place. If you're going to need help to move an acoustic piano, put the piano bench where it can serve as a placeholder for the piano.

- **Place the sound board on the opposite side of the stage from an acoustic piano**

Acoustic pianos are loud. In a small venue, the piano may not need to be amplified at all and it will establish the sound level for all the other musicians and performers. If the band includes an electric piano, you can put the sound board anywhere that's convenient. But if you have an acoustic piano, it's a lot easier to get a good mix without having the full-throated roar of a piano just a few feet away.

- **Put chairs in place for musicians**

I usually start setting up for the musicians by clearing the stage area and putting chairs in place. Remember—musicians need to be able to see each other. Imagine you're a band member and make sure you can see all your mates.

I try to visualize where each person will be sitting or standing with which instruments. Then I move chairs around to provide the room they will need—including room for music stands, additional instruments, personal monitors and other gear.

When I set up a microphone stand for a musician, I may sit in the chair and adjust the stand for the imaginary instrument.

If you can make a table available in the "behind the band" area somewhere, the musicians will appreciate it and use it for instrument cases and accessories.

14 The Incompleat Sound Operator

- **Do NOT put chairs in place for open bands**

If you work with open bands, I strongly encourage you NOT to put chairs on the stage before you run microphone cables, connect and test your entire setup—microphone lines, speakers and monitors. Your work will go much more quickly if you don't have to navigate around chairs and the musicians who will inevitably start occupying them as soon as they are in place. See the chapter on Big Band Sound.

- **Let there be light, cool air or heat**

If your stage is dark, can you provide music stand lights? At least have some on hand? With the LED lights available today, that's easy to do. Fans? Heaters? Set them up or think about where they will go and plan how you will power them.

- **Set up microphone stands for easy adjustment.**

On microphone stands with adjustment screws, turn the booms and tighten them so that all the adjustments are easy for the musicians to reach. I assume that the musicians are right-handed. Tighten—but don't over tighten—the adjustments. For microphone stands with tripod bases: The bottom end of the vertical column should be locked "up" in the base. It should not touch the floor.

- **Prepare an area for the caller/performers**

Most dances have a default location for the caller's microphone. See if you can find a small table and put it nearby. Odds are good it will be used. Other performers—singers and instrumentalists? You've already spoken to them in advance and know where microphones and monitors should be placed.

- **Accommodate the caller/musician**

At dances where the caller is also a musician, contact him/her in advance and talk about how s/he would like to set up. Take the initiative and smooth out the setup.

- **Set out an extra microphone stand and microphone**

If you've got the equipment available, set up an extra microphone stand and microphone off on the side or in the back, just in case. A musician may forget his/her direct. Someone may bring an extra instrument. A musician may invite a friend to sit in with band. Setting up an extra microphone is easy when you're getting the equipment out. It becomes a rush job when it's five minutes before the dance is supposed to start and your unused gear has been set aside. If you've got a microphone setup on standby, you can accommodate a change easily and efficiently. If it's not used, it's no big deal.

I will also leave out one or two extra microphone cables in easy reach; for that extra microphone setup or "just in case."

Look everything over. Adjust chairs. Visualize "your" band playing great music.

Run Cables in the Correct Order
Yes, yes, yes. It matters.

AC Power First: Lay out your AC power cables first. Ideally, there should be only one or, at most, two heavy duty power lines coming from stage outlets to the area where the band is set up. Do everything you can to minimize the number of trip hazards by using power strips and daisy chaining extensions. You seldom need to move AC power cables in the course of an event. Let these cables be the bottom layer of your cabling.

◀ **Tablet era AC power needs**

It is safe to assume these days that some musicians will need AC power for a tablet, effects box, personal monitor or some other accessory. Many musicians use tablets (iPads, Android tablets, phones, etc.) to store their music. Rhythm players or band leaders may have a tablet set up with a metronome or live bpm (beats per minute) app. If you have a keyboard in the band, you need power for it and probably an additional outlet for a personal keyboard monitor. Think about adding an AC power run and power strip to a central location where a band will set up. Chances are good someone will use it.

Speakers & Monitors Next: This is another set of cables that won't move around very much as you complete your setup and run the event. You might have to shift monitors a bit, but generally, after you set up, everything will be pretty much in the same place you expect it to be. Speaker cables are your second layer.

Microphone Cables Last: These are the lines that will most likely need to be shifted. Keep microphone cables on top to make it easier to pick up a line and move it or replace it if necessary. If you are working a concert or something like an open stage, things will move around a lot—mostly microphones and microphone cables. Make them the top layer.

Running Speaker and Microphone Cables

- **Start cable runs at the board**

When running cables, start the cable run at the sound board and leave any extra cable at the speaker or microphone. If there's slack at the speaker, it makes the speaker more trip and tip proof. There's less danger that someone may accidentally catch the cable with a foot and put a dangerous tug on the speaker. For microphones, this procedure lets the musician know exactly how much room there is to maneuver the end of the cable. Coil extra cable neatly. Keep it out from underfoot.

- **Label your microphone cables**

If your group's microphone cables are not labeled, take the initiative. At one venue, we have applied P-Touch labels with a unique two-letter code on each end. Another option is to use rolls of colored plastic electrical tape and put colored bands of tape on each end of each cable. One yellow, one red, one blue, two yellows, two red, two blues, etc.

No matter how carefully you lay out your cables, a time will come when some microphone lines will be switched or moved and you'll be trying to sort out which instrument is on which microphone cable, and you'll be glad to hear someone say—"it's on the line with three red stripes" or "it's got the MG label on it."

The P-Touch labels are a good option since they are easy to read, especially for someone who is color blind. The colored tape bands are probably more durable, but harder to work with when you have a lot of cables to label.

- **Keep cable runs neat**

Try to have all the cables run on the smallest number of paths. You reduce the number of possible trip hazards and make it all look more professional—like the cabling is planned and you know what you're doing. I always make an extra effort to reduce the number to wires that musicians have to step over when they are moving around to take their seats. Just think: Keep it neat.

If you have cables crossing a busy walking path (right at the top of the steps everyone uses to get on stage, for example) cover them with a floor mat or, worst case, tape them down using gaffers tape or duct tape (not recommended, though, because of the messy residue tape leaves on the cables).

◀ **As you connect microphone cables, go left to right**

Standard operating procedure at most dances is to have the caller on the first channel on the board. Then have instruments plugged in in the order that they are placed across the stage as you look at it from the board from the hall toward the band. At a coffeehouse or other type of performance, use the same technique. Identify instruments and voices from left to right so your channels align with what you see on stage. Standard operating procedure gets to be that way because it works pretty well. Use it unless you've got a good reason for doing something different.

A good reason such as?

When I work with an open band I frequently put the keyboard on channel 2—next to the caller on channel 1. The keyboard is a VIP instrument in an open band. I like having it anchored in channel 2.

I use a headset microphone when I call. If the board at a venue has enough channels, I ask the sound operator to put my headset on the last channel on the board. Channel 1 will be for a handheld microphone that I may use, and that will be available for announcements and to back up my wireless headset. If the headset is on the last channel on the board, it's easy for me to find it and make adjustment on the fly if needed.

◀ **Don't mix microphones if you can help it**

Using the same microphones for all musicians simplifies your setup and may help avoid feedback problems. In the traditional dance world, Shure SM57 microphones are the default instrument microphone. The Shure SM58 is a rugged, generally accepted vocal microphone of choice. Unless you have a good reason for doing something different, keep it simple. If you have enough of one kind of microphone to meet your needs, stick with it. Dynamic microphones (the ones without batteries or phantom power) are a little easier to work with than condenser microphones. Using identical, dynamic microphones on stage makes life easier.

As you get more experience and learn home for other sound operators, you may determine that one microphone is better for some applications than another. That time will come. For now, keep it simple.

◄ Put foam windscreens on your SM57 microphones

The Shure SM57 is a great microphone at its price, but it is very susceptible to wind noise and "popping."

Note: Popping is a loud, distorted sound from a microphone caused by the additional air from a person's mouth when saying or singing words that begin with "plosive" consonants. The letter P is notorious.

Get foam windscreens for your SM57s. Benefits: prevents wind noise and popping; dampens loud noises when the mic is accidentally bumped by a musician and protects the microphone during setup and when it's being moved around on stage. Additionally, the foam will protect an instrument if it is accidentally bumped against a microphone.

Extra additional benefit: An SM57 (with foam windscreen) makes a very good caller/vocal microphone. I prefer it to an SM58 personally and use a windscreen-equipped SM57 for my vocal microphone in my small, personal sound system that I use for dance party gigs. Shure makes a foam windscreen accessory that locks onto the microphone. It's not very expensive and worth the money.

◄ Check it all out

When you have all the equipment connected, check to be sure that everything is actually working. Power up all the components in the system. Ask someone to tap on each microphone or speak into it.

The process will vary from here, based on your system. All you are looking for at this point is proof that your microphone works, your cable is good and you have good connections.

Setting Up 19

If you have a signal light or meters on your board, you may just need to raise the channel gain while your helper is testing. If you have a headphone jack and an engineer's monitor (more about that in the section on gain structure) you may use those tools. Or you may need to raise your channel and master faders/knobs and turn up the master gain until you hear your helper.

Once you see your visual indicator or hear the tapping or counting, you know you have a good connection. Now you know your microphone lines are OK.

If you have XLR cables lines set out for instruments that will "plug in," test them by temporarily connecting a microphone and tapping/counting.

Some sound operators recommend getting the channel sound coming through a speaker and wiggling the microphone connection at the board to be sure the connector is fully inserted and to be sure there is no "crackling" caused by a loose or defective connector.

No sound? Start troubleshooting. Think about the 'signal chain' diagram on page 10.

First, make sure you are testing the correct microphone line on the correct channel. If you have that right, check your connections. The most likely cause of the problem will be a cable that is not fully plugged in—just a bit short of the full "snap."

Occasionally you'll find a bad cable or a cable with an intermittent fault or a bad connector. If you identify a problem cable, do two things: 1) Mark it boldly with tape or a repair tag and 2) DO NOT put it back with your unused gear and cables. Take it out of service. Don't leave it to be the next sound operator's problem. See the "if it's bad" point below.

Don't forget to check your speaker connections, too. Be sure you can hear sound from both main speakers (assuming you use two speakers) and from each monitor speaker. I've been at dances where—two or three dances into the evening—someone noticed that one of the two main speakers wasn't working. Also, if you are working with powered speakers, be sure to check the gain/volume settings on the speakers. Under most circumstances, they should be set at the same level.

◄ Label the channels

Have a way to label each channel on your mixing board with lettering large enough that you can read it quickly and easily. I use 3/4-inch wide, white artists' tape purchased at art supply stores. It's more expensive than alternatives like low tack masking tape, but worth it. It's easy to write on with a felt tip pen and comes off without leaving a sticky, adhesive residue. I attach the roll of tape to the board with two, long electrical cable ties. That makes sure the label tape doesn't go walking away.

◄ If it's bad, label it and put it aside or throw it out

Whenever you have a cable (or any other piece of equipment) that doesn't work, be sure to mark it as defective with a big, bold label. If it's a speaker or some other component and you had to 'jiggle" it to get it to work, label it and notify the keeper of sound equipment. (You have my permission to use some of that expensive artist's tape you got for channel labels if you don't have any repair tags.)

If it's a cable, my suggestion is to remove it from service and buy a new one to replace it unless you have a confident sound technician who can repair it. Whatever you do, don't put it back with the spare gear, unmarked, for someone else to have the same problem you had.

Do you have a "repair bin" or a sound guru who can do repairs? Figure out a way to get defective or questionable equipment out of your equipment inventory and into some process for evaluation, repair or disposal.

Microphones and 'Directs'

There are lots and lots of different types of microphones. If your group has several different types, take some time to get to know them. Cable each different type up and try it out. Move to a place where you can hear your main speakers clearly. Talk into the microphone and get a sense of what it sounds like. What happens if you are very close to it? Far away? Off to the side? When you say words starting with "p" and "t"? Try speaking directly into it, and then from the side, to see its range of sensitivity. The microphones you will find most frequently are Shure SM57s and SM58s. The SM57s provide crisp sound reproduction but are sensitive to "popping." The Shure SM58 is designed for vocal use (it has an internal wind screen) but also has a strong "proximity effect" (when you get closer to it, the sound has more bass in it).

◀ **Play with your Microphones**

Because you arrive early and have things set up well in advance, you have the opportunity to experiment with microphone placement. Get a musician to play something (any old tune) and try moving the microphone (or your ear) to different places around the instrument. Find the "sweet spot" for different instruments.

Experiment with your microphones so you have a sense of how they work in different locations. The positioning of a microphone can be the key to solving many problems. For example, you might find your levels are extremely high for an instrument and wonder what's going on. Look at the player. He or she may be six or eight or ten inches away from the microphone. You may need to help and get them to move the microphone closer.

Warning: Remember, not every musician or caller is familiar or really good at working with a microphone. Some callers—even very experienced callers—will hold a microphone an inch or two from their mouths when teaching a dance and then hold it twice as far away when actually calling the dance (or vice versa). The changes affect both the volume and the quality of the sound. You have to find ways to compensate which may include coaching the caller or riding the sound board throughout the evening.

Microphone Placement for Musicians

The following suggestions are good for a "getting started" setup. (Disclaimer: Every sound professional sound tech will disagree with some, most or all of my suggestions here. There are lots of opinions on this.) Most of the time, you'll be working with musicians who have been working with microphones and they will have a pretty good idea how they want the microphone to be placed. Feel free to suggest other options if you have strong feelings, but don't argue. Your job is to serve the music and that means serving the musician.

Fiddles and Violas

Above the instrument, pointing down toward the upper F-holes, as close as the musician is comfortable having it.

Cellos and Basses

Low, below the musician's hands and bowing area, aimed toward the bridge more than the F-holes. Cellos are tricky and you will tend to get more bass than you want/need. Look for a placement that gives you the best mid-frequency and high notes.

With a bass, the notes will carry. General vicinity is fine. Some bass players like to wrap a microphone in cloth and tuck it under the strings of the bass below the bridge of the instrument.

Guitars

Below and a bit toward the neck from the sound hole. Close to the instrument but not directly over the sound hole. Right over the sound hole produces a more boomy, bass sound. Away from it, you get full sound plus some of the pick/string sound you want.

Mandolins and Ukuleles

As close as you can get comfortably in front of the instrument.

Banjos

I use similar placement to a guitar. The sound is broad and diffuse, so let the musician be your guide.

Hammered Dulcimer

My preferred mic location is directly underneath the instrument. I've done test with headphones, and the percussive hammer and string sounds comes though fine. Hammer dulcimer players will likely resist such placement. So do go for as close as possible to the instrument from up above at the high end of the instrument while clear of the player's hammers. The challenge with mic placement for this instrument comes with monitors. The player is standing and sometimes wants more monitor volume. And the microphone, pointing down toward the hammered dulcimer, is also pointed toward the monitors. We've used piezoelectric contact microphones successfully with hammered dulcimers.

My favorite hammered dulcimer players have built pick-ups into their instruments.

Accordions

If you have two microphones, put one on the bass side and one on the treble side. Only one microphone? Favor the treble side where the melody is played.

Concertinas

Two microphones, put one on each side directed in the direction of the player's hands. One microphone: position it a little above and in front of the instrument.

Flutes, Whistles and Recorders

Above the instrument near the embouchure (where musician blows to play a flute) or the labium (sound-producing lip on a recorder or whistle).

Clarinet

I like to put the microphone above the instrument about a foot away from the reed. Some clarinetists prefer having the mic aimed directly up the barrel of the instrument.

Saxophone

Anywhere in the general vicinity of the instrument.

Pianos

Uprights, I like to mic from behind. You can use two microphones, one toward the bass end and one toward the treble. If using one microphone, favor the treble side a bit. For a grand piano, I'd keep the microphones below the instrument for a dance band. If you are working a jazz/classical gig, you'll want the lid raised and mic above the strings.

Percussion

For a bodhran and other hand drums, let the musician be your guide. In general, percussion sound carries quite well. You probably will not need a lot of amplification and you likely will not have it in the monitor mix, so you'll have lots of flexibility. Consult the drummer, the Internet and other sound resources for mic placement for a drum kit. I've never had the pleasure.

Going 'Direct'

More and more musicians have their own microphones or instrument pick-ups (small, clip-on microphones or, especially in acoustic guitars, a built-in microphone). They won't want a microphone. They will ask to "plug in." In general, musicians and sound people will refer to this as going "direct" to the board. Typically, all you have to do is provide them with a microphone cable that has a female XLR connector.

The two most common scenarios you will meet are:

The musician has a "direct box."

All you need to do is provide an XLR microphone cable. The musician will plug the cable into his/her direct box. and you'll be ready to go. Maybe.

Things that can go wrong:

- Direct box switches and level controls. Sometimes a musician will forget to turn the switch on or the sound level (gain) control on the box will be set very low or very high. If you don't get a level right away as you start checking the level, turn your channel settings back down the zero and ask the musician to check his/her switches and gain control.

- Batteries. Some direct boxes have batteries. Batteries can go dead. That's another point to check if you're not getting good sound from the direct.

- Phantom Power. Some equipment requires power from the board known as "phantom power." This is usually a 48 volt DC current. Many condenser microphones need phantom power and it is also required for many "active" DI boxes.

Setting Up 25

- Connections. Cables, switches and plugs have the potential for simple mechanical failure. Springs get less springy or tiny wires break after repeated use. If you've checked everything else, try setting channel levels very low and wiggling the connections and wires, to see if you can get a signal.

When all else fails and it's getting close to time for a sound check and to start the dance/event: You have a spare microphone set up. Get it in place and ask the musician to use it until you have time to sort out the electrical problems.

The musician plugs in with a quarter-inch instrument cable.

This is how a keyboard or electric guitar/bass might be connected to the sound board, especially if no DI box is available.

Usually the instrument will have a gain (volume) control, so be aware of that when you start setting levels. Sometimes an instrument like a keyboard will send such a strong signal that you get a peak signal with the channel gain set at zero. Some boards have a "pad" button/setting that allows you to reduce the line input by 10 db. This is the time you want to use it. On some boards the "pad" function must be engaged to use a quarter-inch input.

◄ **Differences between speaker and instrument cables**

Use labels or special color-coded markings—something—to help keep speaker cables separate from your instrument cables They are different. Instrument cables and speaker cables look pretty much the same. Both have quarter-inch plug connectors on each end, unless you are using speaker cables with speakON connectors (a good idea if your equipment supports them). They look the same, but inside, they are different.

An instrument cable has one or two thin wires and a braided, flexible wire mesh shield. A speaker cable has two thicker wires inside and no shielding.

Instrument cables are designed to carry a low voltage signal and protect it from electrical interference—noise, static, buzzing and other sounds that can be caused by electric motors, fluorescent lights or other electrical equipment.

Speaker cables have bigger wires inside and are designed to carry a larger electrical current from the amplifier to the speakers.

If you use speaker cables for instruments: you may have outside interference in your sound. If you use instrument cables for speakers, the wires may melt and bad things could happen to your amplifier, according to cable manufacturers.

In the relatively quiet world of acoustic music (vs. festivals and arena rock shows), you might be able to get away with using these cables interchangeably. But if you're going to do the job, why not do it right?

◄ **When time is short, get the directs set first**

Give direct connections priority when cabling and checking musicians. Plain old microphones are easier to set on the fly.

Wireless Microphones

All wireless microphones have two key components: a transmitter that is physically connected to a microphone or instrument pickup; and a receiver that is connected to your mixer.

A wireless handheld microphone has a transmitter built into the microphone body. Headset microphones and instrument microphones typically include a "body pack" of some sort. The body pack contains the transmitter, on/off and mute switches and the battery or batteries. The headset or instrument will have a wired connection to the body pack.

The receiver is a small, AC-powered electronics box. It may have an antenna or dual antenna; it may have a gain control and it will have a input(s) for a microphone cable: XLR, quarter-inch or both.

When you are working with a group's standard setup—your house sound system—any wireless gear will have been tested in regular use and have a track record. When the gear is connected properly, it will work. You should only have one thing to worry about—having spare batteries.

If a musician, caller or other performer brings wireless gear to use, you may run into other issues—most notably electrical or radio frequency interference. The radio frequency used by wireless sound equipment has been changed at least twice over the last two

Setting Up 27

decades. Gear that works fine in one place may fail in others. My Shure headset system works well almost everywhere, but not in the Spanish Ballroom at Glen Echo, Maryland, and the Irish Center in Philadelphia. Since you will already have communicated with the performer and know that wireless gear will be involved, you can be prepared to get it set up early and tested.

◄ **Batteries, batteries, batteries**

Be sure you have spares for the gear you know you will be using. Typically, you will need AA or 9-volt cells. I've tried to encourage using rechargeable batteries, but in an environment where there are several operators of the same equipment, I have not had success.

◄ **Set up and test 'foreign' gear early**

Advance communication pays off. Get the wireless gear set up and tested as soon as possible. Some interference problems can be overcome by changing the channel settings on wireless gear that has multi-channel capability; but you need time to work that kind of thing out.

◄ **Have a wired microphone backup**

In my role as a dance caller, I frequently use a wireless headset. And I always ask to have a wired handheld microphone set up. It's a backup for my headset and it's available for announcements.

Working with Artists

I'm in awe of people who can produce the tens of thousands of notes that become the music that propels a dance or performance. It seems to me that the role of the sound person is to make life as easy as possible for these talented folks; to have as many pieces in place when they arrive so that they feel as welcome and respected as they deserve to feel.

The ideal situation as I see it, is to have musicians arrive and be able to sit down and have everything in place. All you have to do is your basic level setting. You're ready to start—even if a band member arrives five minutes before the event is scheduled to start. If you can manage to have everything ready, you're creating the right environment for a successful evening.

It's been my experience that if you work to be a good host, the musicians you work with will recognize your efforts and respect you in return.

- **Don't take it personally**

When you encounter a musician who behaves disrespectfully, speaks harshly, or embarrasses you in some way, don't take it personally. There may be other things going on that you don't know about. Just do the best you can and "let it go." If a musician really misbehaves, don't agree to do sound the next time that person plays. Reality: There are prima donnas. More often, though, a musician may be a little insecure or nervous, and say something he or she didn't really mean. You may be able to learn their quirks, gain their confidence and adapt to keep things harmonious. That's the best thing to do. But if an individual or a band a big problem for you, avoid them in the future.

- **Pay attention to the musicians**

If you stick close to the sound board and regularly look at the musicians during the first half hour of the evening, your musicians will appreciate you. If you see a musician making eye contact with you, run (don't walk) over and find out about the concern.

◂ Watch the musician as you make adjustments

The most frequent request you will get from a musician will be for a change in the volume of one or another instrument in the monitors. Once you know what the musician wants, go back to the board, find the control, and *slowly* adjust the level. You'll get a clear signal from the musician (probably a nod and maybe a smile) when the settings are right.

Gain Structure

Initial Settings to Help You Get the Volume Right

Gain structure involves getting the gain/volume settings on all channel and master faders (or knobs) to an optimum basic setting. You will want to have everything set to the levels the system components were engineered to support while leaving yourself room to adjust volume up or down as necessary when zeroing in on your "final" mix.

The following section is the most technical part of this guide. It is also extremely generic. It's written with the intent of being applicable to just about any sound setup, from a four-channel powered mixer to a 64-channel digital board. And since the systems people work with may have powered mixers (mixing board and power amplifiers in one unit), separate power amplifiers or a mixing board feeding powered speakers, this section will assume that you will make adjustments as needed to balance your mixer settings with the amplifier settings that drive your speakers.

Terminology

A few things that need to be defined.

Unity gain: This is a biggie. Unity gain is a setting at which the signal coming into a component in the system is equal to the signal going out of the component. The components that we want to focus on are the faders on the board of the mixer—the individual channel faders and the master faders. Typically, there will be marking on each fader's scale that shows a with an obvious focal point (0db or 0 or U as shown in the photo on the right), about two thirds of the way up from the bottom of the faders' adjustment track. That specially marked spot is the "unity gain" position.

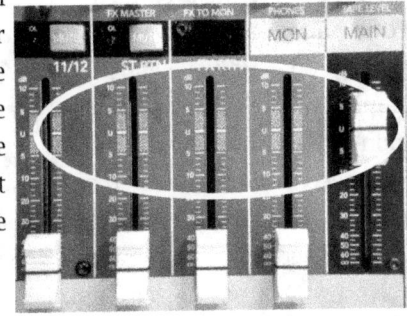

If you have a board with dials/knobs instead of faders, or you are working on a sub mix (monitor mix for example) controlled by dials/knobs, you may not have a scale showing the unity gain setting. An educated guess will be to turn the dial about two-thirds of the way up. If a zero setting—knob turned fully counterclockwise—is at seven o'clock on an analog clock face—a full on (clockwise) is at about five o'clock—then unity gain will be around two or two thirty.

Gain: Most mixing boards will have a single knob near the top of the controls for each channel that is used to reduce or amplify the signal coming from a microphone, instrument pick-up (microphone), keyboard, or other source. Some engineers may refer to this as a pre-amp control. Small powered mixers probably will not have a gain knob. The only volume control you will have for each channel will be a single knob or slider.

Meters: Some old boards may have a needle-type meter to show signal strength; newer boards may have a series of LEDs; green and the bottom going about two thirds of the way up, then yellow and then one or two red LEDs at the top. The LEDs are easy to interpret. Green is good. When the signal is making the green LEDs dance, occasionally making a yellow light glow, but is not making the red LEDs flash, you are in a good place.

Signal light: Some boards have a green LED that starts lighting up when the channel gets a signal. It will stop flashing—just show solid green—when the signal gets too strong. A nicely blinking green "signal" light is your friend.

Clip light: A "clip" light is a danger signal. It flashes when the signal is too strong and the sound may "clip" and be distorted. Flashing red lights are much to be avoided.

Power amps and powered speakers: Some systems may have a mixing board that sends its low power output to a power amplifier. The signal then goes from the amplifier to "passive" (non-powered) speakers.

In recent years, powered speakers have become more popular. In these systems, the low powered signal is sent from the mixing board directly to the powered speaker. Powered speakers have their own, built-in power amplifier.

Gain Structure

With either of these setups, you can make adjustments to the overall sound level at the amplifier or speaker.

Powered mixers: Powered mixers have a power amplifier (or two or more) built into the mixer—the mixing board and power amp are all one unit. With a powered mixer, all your gain/volume adjustments will be made on the powered mixer. You may have three or two controls. You may have a master gain for each channel as well as a fader and board master fader. Or you may have one gain/volume control for each channel and one master control for the board. Most small groups/venues have a powered mixer.

Pre-fade and post-fade: These terms refer to where, in its path through the board, a signal branches off from the eventual mix that is sent to the speakers. A "pre-fade" signal is not affected by the adjustments you make on a channel fader. In general, if you are sending a signal to a musician's monitor and you want it to be consistent after you set it (not change if you adjust the volume of the channel in the house mix), it should be "pre-fade." A "post-fade" signal will change as you adjust the house mix. If you wanted an accurate recording of the music as it was heard in the house, you would want a post-fade signal.

PFL: The acronym for "pre-fade listen." Many mixing boards will have a headphone jack, headphone volume control, and a pre-fade listen button for each channel. Pushing the PFL button sends the signal on the channel to the headphone jack. It allows the operator to hear the sound from that channel, even if all the other faders/knobs on the board are fully off.

Setting Up Your Gain Structure

Everything is ready: Your board and related components are in place, speakers and monitors are set up, microphones are in place and cabling is complete. You are ready to start building your mix.

Phase 1: Put the System in Neutral

Zero out the board. Set gain control for each channel to zero. Set all the EQ controls to neutral. Set all monitor and auxiliary channel controls to zero. Set faders and master faders to zero/full off.

Phase 2: Get "In the Ballpark" with All Channels

As musicians/instruments become available, have them play and adjust the channel gain to a "good" level. How do you know what a "good" level is?

The board I use most frequently has a green "signal" LED. I look for it to be illuminated—going on and off as the musician plays—but not a solid green. A happy, bouncing green signal on this board says "good" to me. Similarly, you can use a board meter to get a "good" gain level for each channel.

If you have PFL (pre-fade listen) on your board, you can check each channel using headphones or (highly recommended) an engineer's monitor.

Another option: turn up your master fader to unity gain and adjust your channel fader and channel gain to hear the music in the speakers. Bring it up to a volume that seems close to what it will be for the performance. You'll make finer adjustments later.

As you identify and check each channel, update the labels with information anyone can use to identify who/what is on each channel. If musicians are "plugging in," try to get to them early in the process. Direct inputs are the most frequent sources of problems when you're at this point in the set up.

Warning: Any time a musician "plugs in" with a direct connection, be sure the channel gain is fully off. Some direct signals can be very strong and bad things can result if your gain is turned up. Have the musician play and start bringing the gain up slowly until you see you are getting a signal. *If your system has a mute switch for each channel, use it to "kill" the channel any time a musician plugs in or unplugs an instrument.*

What if there is no signal on the "direct" channel. Start going down the troubleshooting tree. Do you have the correct channel? Did the musician connect to the correct cable? Is there a switch the musician needs to turn on? Do you need phantom power? Is there a battery that may need to be replaced? Recheck all physical connections. In the worst-case scenario, you have a microphone stand and microphone set up on standby.

A Shout-out for Visual Indicators
Note the benefit of using signal lights or board meters at this point

in the set up. If a visual indicator is available, it makes getting to this first, very rough mix, a quick and reliable process. You can get your basic gain structure in place without even having the speakers on. When working with large bands, you can identify players, get them to play when they are set up and on a microphone or plugged in, and you can have a basic mix ready to go entirely by sight.

Phase 3: Balancing the Sound in the Monitors and the House

When you have a good signal on all the channels, you can move your channel and master faders up to unity gain and you should have a good start on your final mix.

What if the sound is too loud when you have channels/masters at unity? If your system has a separate amplifier or powered speakers, you need to reduce the gain settings for them. Turn down the amp or powered speaker(s) until you can have the mixing board faders close to unity. Or, if you find yourself close to the top of the adjustment range for the faders, you may need to turn the amp/powered speakers up.

No separate amp/powered speakers?

If you're working with a powered mixer, set your channel faders at unity gain and adjust the use the master faders as necessary to control volume. The masters may not be near unity gain but you will be making fewer adjustments to them than to your channel faders/knobs. You want to leave yourself room for adjustments on the channels.

Digital Mixers—An Aside

You've seen sound operators walking around a room or going on stage with a tablet or mobile phone and making adjustments to the mix. The cool factor for this kind of technology is through the roof.

If you are a pro engineer or you are putting together a rig that you and only you will be using, you'll find digital boards have come down in price to a point where they are quite competitive with the kind of "old-fashioned" analog equipment we've been describing in this book. The equipment works well, it's reasonably affordable and gives you amazing control and flexibility. So, go for it.

However, if you are working with a group that will have different people setting up and running sound, I'd strongly encourage you to stick with that "old-fashioned" analog gear for several very good reasons. Analog gear is time tested, well-engineered and works very, very well. Faders. knobs and dials are easy to understand. One analog board works very much like another. With digital gear, every manufacturer has a different kind of interface.

Digital gear has a steep learning curve for the operators. And you can run into all kinds of issues you simply don't have with analog gear such as: knowing the passwords for a control pad; synchronizing a remote control with the main board and Wi-Fi connectivity for the system.

Gain structure complete?

You have a good signal on all the channels you are using. You have a good sound level in the hall, with your channel and master faders at, or close to, unity gain.

At this point, if the setup is running late, you can start the dance/performance. Run up your channel faders and bring the monitor controls up about halfway—watching the musicians carefully—and in a worst-case scenario you can declare sound is "ready to go" and fix/adjust the final mixes for mains and monitors on the fly.

You are ready for the sound check.

PFL and an Engineer's Monitor

While working at a large dance festival, I saw that several of the engineers used personal monitors rather than headphones. I tried it and I think it's a much better way to access your PFL functions.

Connect a small powered monitor to the headphone jack on your board. Turn up the headphone volume a little bit. Then use the volume control for the channel you are using on your personal monitor to hear what's coming through the PFL/headphone out selection.

Advantages? I find it much easier than working with headphones—getting them on and off and figuring where to put them. The personal monitor is faster and easier once you have the hang of it. And, if your board has the ability to route different mixes to your headphone jack, it gives you a quick and easy way to do things like listen to your monitor mix without going up on stage. Sure, you can use headphones. But, in a live sound environment, many professional engineers use personal monitors instead. If you can, go with the pros.

Sound Check

Which Comes First, Monitors or Mains?

You have all the musicians in place and ready to play for a sound check. Should you start with the monitor mix that the musicians are hearing on stage? Or should you start with the sound out in the house?

My approach, over the years, has been to focus on the house mix—with monitors off. Comments from sound operators who have reviewed this text in draft form have made me change my approach a little bit.

There is a good case to be made for focusing on the monitor mix first. You make the musicians more comfortable and they may play a bit differently when they are hearing themselves well, which might change your house sound settings if you have already dialed them in.

If you have time, all the musicians are present and setup is well in hand, it seems to me that starting with the monitor mix makes sense. A quick conversation with the musicians about the mix will give you the big picture. Should you include a bass in the monitors (usually, "no") and what about the piano? "Just a bit." If you have done a good job setting the gain for individual instruments, you should be able to get a mix set quickly, and the musicians will be playing—warming up. When you turn to the house mix, you'll be closer to performance sound levels.

However, if you are running late for any reason and/or musicians are late—focus on the house sound. Be sure you have a "good enough" mix to get the event started on time. Then tweak the monitor mix as the event progresses.

Final thoughts: If you've done a good job setting up the gain structure and have a nice "just about right" signal on each channel you are using, you will be able to deliver a good "just about right" mix to the monitors and the main speakers at the same time. Get the basic structure right and everything else will be easy, making minor adjustments.

Sound Check: Monitor Mix

If you have not done a separate sound check for monitors, you can bring up a basic monitor mix that reflects your basic gain structure. Bring the monitor channel and master levels up to a low level. When the band stops, ask the musicians about the monitor settings. Make adjustments. If the request is for "more fiddle" or "less piano," adjust the monitor control for that instrument's channel. Almost always: Do not include the caller in a dance band monitor mix.

◄ **Caution:** Do not change your channel gain (preamp) control settings after you start setting monitor levels. On most boards, changing the gain on a channel will also change the monitor volume level.

◄ **More Caution:** On rare occasions, you may need to change channel gain control settings during the dance. Why? Most musicians will play louder as they "get into" the dance. Some will play significantly louder causing the clipping indicators to flash. Reduce the channel gain control. Be aware that you are changing the monitor monitor mix. Watch/listen closely to be sure the musicians are still happy

Sound Check: House Sound

Having a sound check before the dance/performance starts is very important. And it's not only for the sound operator. A sound check gives the musicians a chance to warm up, begin to relax on stage, and generally start getting into their music. If they are new to the venue, they begin to get a feeling for the sound there.

NOTE: In some venues, stage house acoustics can have a major impact on the sound in a hall. The stage house can serve as a

Gain Structure 39

resonating chamber and act like an additional speaker, sending sound from the stage (especially the monitor mix) into the hall. Something to be aware of.

Ask the band to play a tune. Set all the faders for the channels in use and the masters at unity gain. Start the fine-tuning process in the house. Start making the minor changes you'll need to get a good balance among the instruments.

Use the individual channel faders to modify the mix.

Use the "master" fader(s) to adjust the overall volume.

The mix you get for a sound check may change significantly as the musicians warm up, begin digging into their instruments, and more dancers enter the hall, or the seating space fills up.

Fine Tuning the Mix

The most important element of getting a sound mix right is the volume of each channel/instrument. The second most important thing is the volume of each channel/instrument. The third most important element is the volume, etc.

Most sound boards have a variety of controls you can use to shape the sound: Adjusting the bass, mid-range and high frequencies. But the primary control—the most important one to focus on as you are starting up—is volume.

In a good mix:

- Lead instruments such as fiddle and mandolin are "in front"—melody lines are clear and distinct
- Rhythm instruments are in the musical background—but are solid and distinct when you listen specifically for them
- Caller/vocal is crisp and clear—a little bit "in front" of the band mix, but not too much

◄ **Extra Ears**

If you are feeling at all unsure about the mix, ask people around you who you respect what they think. Every person hears everything just a little bit differently. "Extra ears" will help you build a better mix.

Even if you are NOT insecure or worried about the mix, you should still solicit feedback. Ask people whose "ears" you respect for their thoughts.

It's not ordinary human nature for people to come to you and make suggestions unless the problem is really egregious. Your friends aren't going to want appear to be critical. But if you invite them to offer comments, you may find that they have valuable insights or suggestions. You may learn something. Make yourself open and you will be rewarded.

◀ **Responding to suggestions**

When someone (anyone) comes to you and suggests a change in the sound mix, listen to him/her. Go over to the board and try out the suggestion (even if you only give it a minimal adjustment). And *be sure to thank the person* for the suggestion. It may be an excellent thought or it may not be, but if you respond positively, you'll help add to the positive atmosphere for the event.

◀ **Walk the hall**

Be sure to walk all around the room to listen to the mix. Move all around. Stop and listen: up font near the speakers, right in the middle of the hall, toward the back, on the sides. It may sound good up front, but be muddy in back. Check the spot right in front of the band between the speakers. I typically angle the speakers in toward the center of the room just a bit to cover that spot. Other engineers sometimes set up a small powered "fill" speaker there.

Sound Equalization (EQ)
Shaping Sound: Frequency Asked Questions

Working with equalization or "EQ" is where you enter the "dark arts" of sound technology. As you get more deeply involved in sound reinforcement systems and how they work, it's a fascinating subject. Experts can discuss EQ for hours. *For the new sound operator, keep it simple.*

Equalization controls raise or lower the volume of sound in specific frequency ranges.

There are usually two places where equalization takes place.

First: Each channel has EQ controls. Usually they are knobs for low, middle and high frequencies. They divide the sound spectrum into three broad ranges—low, middle and high—and each knob raises or lowers the sound volume for that part of the sound spectrum.

Additionally: Many boards/mixers you encounter will have two sets of multi-band equalizer controls covering between about six and twelve tighter frequency ranges (bands). Some systems you work with may even include a more advanced, dedicated graphic equalizer with twenty-five or more bands; a separate component placed between the mixing board the power amplifiers or powered speakers.

Most systems are set up so that one set of controls affects sound for the house speakers and the other affects sound for the musicians' monitors.

The multi-band equalizers divide the sound spectrum into smaller segments, and give the operator the ability to raise or lower sound volume in the selected frequency range for the entire mix (not individual channels).

The best place to start working with equalization settings is with everything at neutral (or at "house" settings, if your local sound experts have determined what they are).

Take a look at the Sound Frequency Diagram provided on the next page. It shows the ranges of various musical instruments and human voices superimposed on the range—low to high—of

42 THE INCOMPLEAT SOUND OPERATOR

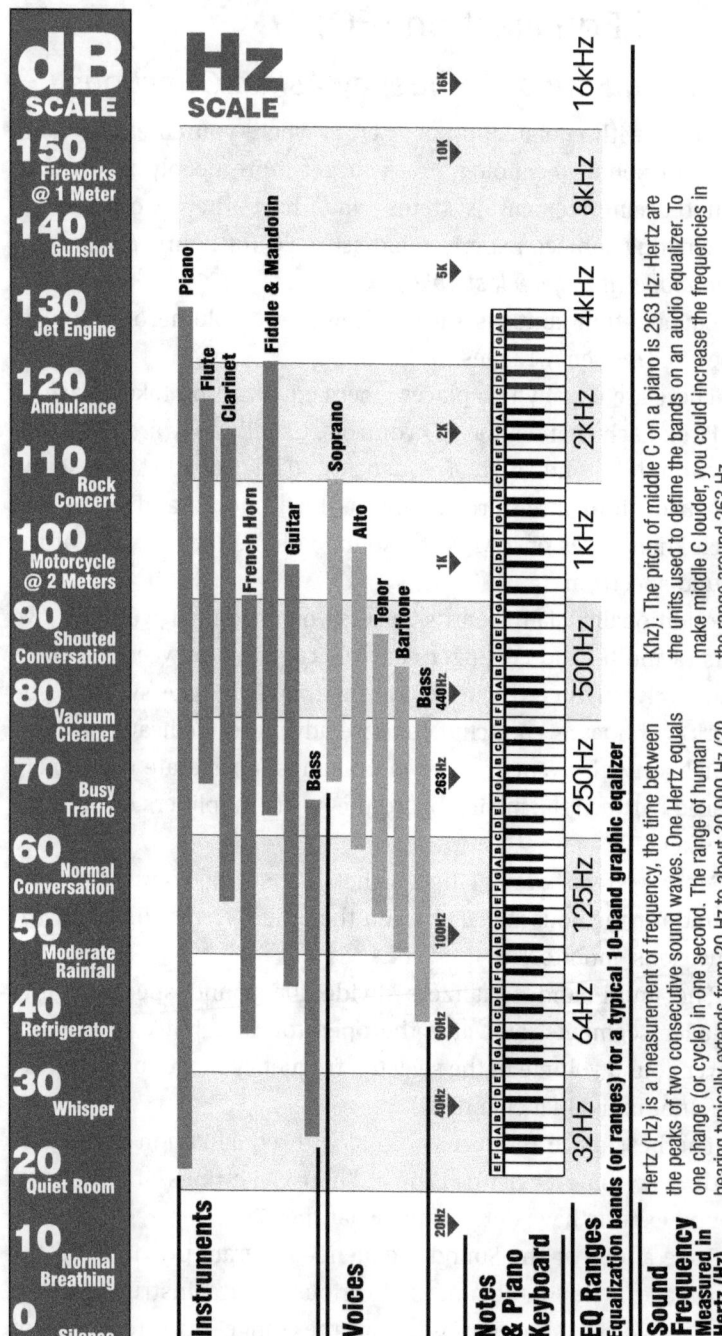

EQ Controls – The Basics

human hearing. The diagram also includes representation of the frequency ranges (bands) of a 10-band graphic equalizer such as you might find on a mixing board or powered mixer. It is helpful in improving your understanding of where the "sound" is relative to the "numbers" (sound frequencies measured in Hertz). It's also helpful to see where male and female voices fit in the picture—to help see how to make callers/vocalists clearer in the mix—and generally to see how much of the sound we most want to hear is concentrated between 100 Hz and 2KHz.

EQ Controls—the Basics

The following diagrams and descriptions are designed to provide new sound operators with a general description of the equalization controls you'll find on a typical sound mixing board. However, every mixing board and powered mixer is different. So take everything that follows as generic, hopefully useful, background information. The following diagrams provide are designed to provide a generic description of what is happening then you work with the kind of equalization (EQ) controls you typically find on each mixer channel.

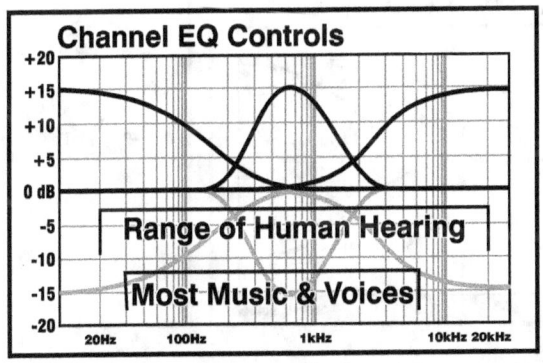
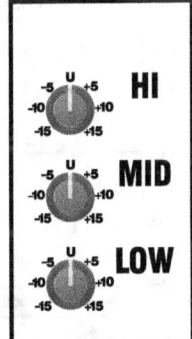

Interpreting these diagrams

In the background grid X-axis, going from left to right, shows a sound frequency spectrum measured in Hertz. One Hertz, named for German physicist Heinrich Hertz, equals one sound wave cycle per second. The average person can hear sound from between 20 Hertz (20 Hz)—the lowest notes played by a pipe organ, up to about 20,000 Hertz (20kHz). However, the highest tuned pitch of musical instruments is around 5,000 Hertz. The y-axis shows the range of adjustment for a typical sound mixing board. The amount of change you can make is measured in decibels (dB). Turning the dB control counterclockwise from its neutral (zero) position, will reduce the sound level for a range of frequencies; turning the knob clockwise will raise the sound level for the affected frequencies. On a board with simple controls such as the HI / MID/ LOW controls shown here, the actual frequencies that will be changed are predetermined by the board manufacturer.

If you leave the dB control at neutral, the sound profile will be the straight line, right to left, at zero. Operators will refer to this as "flat." The dark curved lines in the diagrams represent the maximum and minimum amount of change. Most of the time, the actual settings will be somewhere between the dark line extremes and flat.

Familiarity with how the controls work and the range of sound that they impact will give you a better understanding of how you can use them to "tweak" the mix to make it sound better, or to make corrections to prevent feedback, for example.

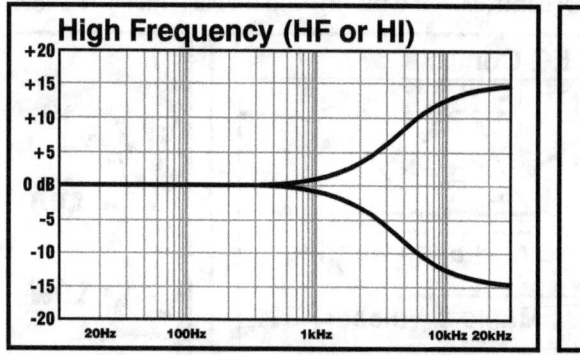
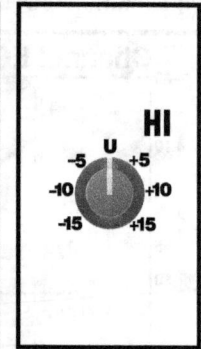

High Frequency (HI) controls allow you increase, or more often reduce, high frequency sound. You might, for example, want to turn down the high frequency control for a fiddle or a penny whistle that is sounding shrill. You might address high, ringing sound that is threatening to become feedback by turning the HI control down. Occasionally, you may want to boost the HI to help make a person's voice crisper, or to get a more vibrant sound from a guitar.

EQ Controls – The Basics

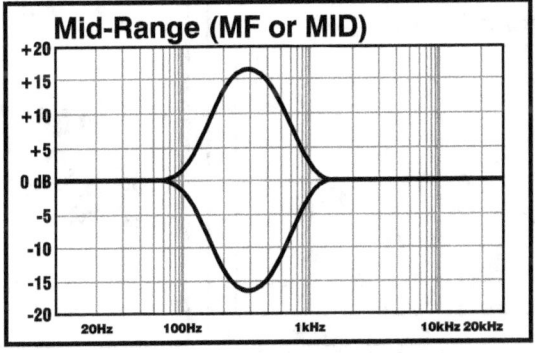
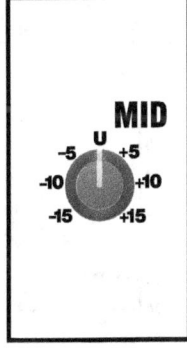

Mid-range EQ occupies the heart of the music and voices you'll be working with. Boosting the MID EQ may be helpful when you want to brighten the sound of a guitar or piano. The MID setting can have a major effect on voices. You'll want to play with the controls and hear the result. With that said, much of the time—especially with musical instruments—the right thing to do is leave the MID alone. Keep the setting flat. If you have a good musician playing on a good microphone, there won't be any reason to change anything.

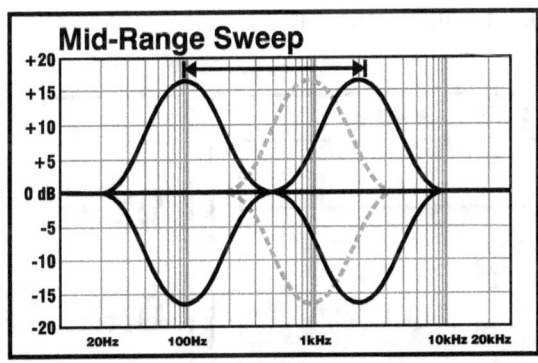
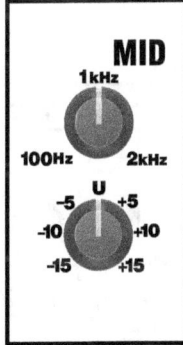

Mid-Range Sweep controls give the Operator the ability to change the frequency that the dB control will affect. In this diagram, the Hz control allows the operator to move the peak of the frequency from a low of 100 Hz to the high of 200 kHz (the heavy black lines). If you leave the Hz control at its neutral setting (twelve o'clock) the frequency range will be centered at 1 kHz (gray dotted line). Many better-quality mixing boards will have one or two pairs of mid-range sweep controls, giving operators much more control over their mid-range EQ settings.

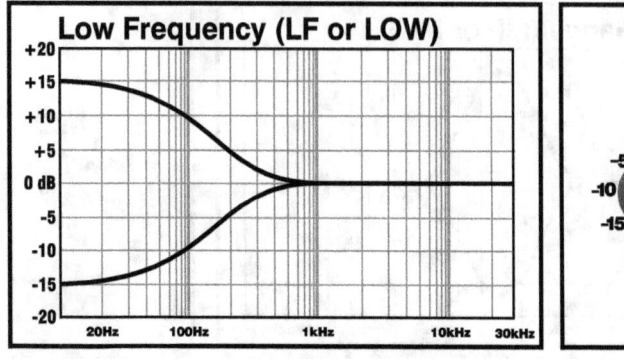 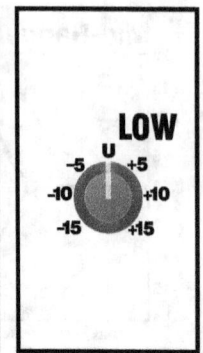

In my experience, the low frequency EQ is used most often, and most of the time, it us used to reduce ("roll off") the bass. Cutting back on the bass can be helpful in getting voices clearer, compensating for the proximity effect in vocal microphones like the SM-58. I roll off the bass most of the time when adjusting sound for a dance caller. You may find it useful to reduce the LOW when working with instruments like keyboards, guitars and even bass fiddles—using it to reduce the "boomyness" of the sound and getting a crisper sound from the instrument.

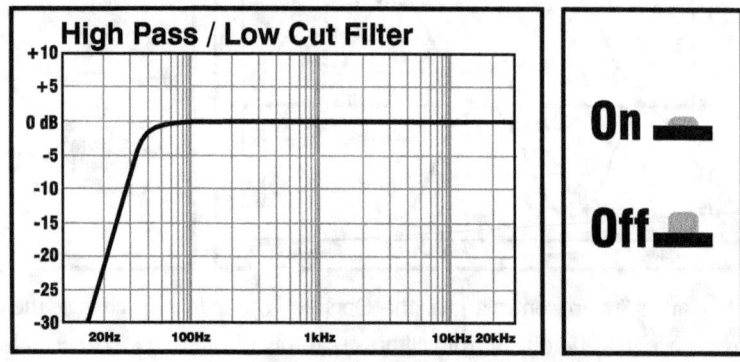

High-Pass / Low Cut Filter. "High-pass" and "low-cut" mean the same thing. Some manufacturers use one term; some use the other. High pass: All of the frequencies above a certain frequency "pass" through unaffected. Sound at lower frequencies is filtered out. Low-cut: Low frequencies (below about 20 Hertz) are filtered out. This filter differs from the rest of the EQ controls covered

EQ Controls – The Basics

here: it's a simple on/off switch. It's designed to prevent unwanted noise from the stage—the sound of a microphone stand being moved, for example, from going into the mix. If your board has High-Pass Filter (HPF) or Low Cut Filter, you should turn it on.

◂ **Play With EQ**

Create opportunities for yourself to experiment with EQ—to make big dramatic changes in settings and hear the results of the changes.

Get a recording of voices and music similar to the ones you are mixing sound for and play it through one of the microphone channels on your board. Experiment with the EQ controls and hear how they affect the playback. There may be opportunities before a dance or concert. You can "play" with EQ for a dance caller during the walkthroughs of dances. Consider doing a sound workshop: Set up your system and have musicians play.

Every time you do sound, it's an opportunity to listen and learn. And frequently what you will learn is that less is more.

Rule Number One: Keep it simple.

◂ **Frequent EQ Adjustments**

The most common changes I make in EQ are:

- Reducing high frequencies in the multi-band EQ settings for the monitors. Whenever you or the musician hear a high ringing that threatens to become feedback, start cutting back on the high frequencies in the monitors.

- Any other changes in EQ the musicians request for sound in the monitors.

- Reducing high frequencies on channels for certain instruments. I frequently find myself turning down the high EQ for some fiddles, whistle, flutes—instruments that live in high frequencies. Sometimes I raise the mid-range for those instruments, too, to try to "warm up" the sound to my taste.

- Reducing very low frequencies for pianos/keyboards and guitars. Increasing mid-range settings for them to get a brighter sound from the keyboards and to hear more "string noise" in acoustic rhythm guitar.

- Adjusting (up or down) all frequencies for dance callers. The goal is clarity. With men, a standard starting point—

especially for men using SM58 microphones—is to reduce the bass significantly and maybe increase the treble (high frequency). Then see if there is a sweet spot in the midrange settings that improves the clarity of the voice. For some women, it's frequently helpful to reduce the treble. Possibly reduce bass. Look for the sweet spot in the mid-range. The rest is up to you and your ear.

- Experiment with dance caller sound
  Vocal clarity is critical when doing sound for dances. Slowly raise and lower EQ settings for the caller during walkthroughs and listen to the changes in the sound. Try exaggerated settings, just to hear what the differences are. Each caller's voice is different. Small adjustments can make major differences in clarity.

Running the Show

Listening and Adjusting During the Dance/Event

◄ **Walk the hall.**

Keep checking. The sound changes as the dance/performance goes on. Move all around the hall: front, back, middle and sides. Stop and and listen to the sound. Is the caller equally clear all around? Is the sound level sufficient in the back? If you stand directly in front of a speaker, is the sound bearable? (It's going to be loud—but is it too loud?)

◄ **Pay attention to the sound volume levels.**

As musicians get warmed up, they start playing louder. As callers get warmed up, they may get louder. Keep an ear on things and listen for these kinds of variations.

◄ **Consider buying (or getting your group to buy) a Decibel (db) meter.**

As referenced earlier in my notes about tools, a dB meter can help you maintain consistency from one event to the next. Every band and every venue will be different, but after a while, you can find what overall levels seem "right" to most people, and then measure and match your standard setting consistently.

◄ **Ask musicians what they think.**
  Listen to what they say.

Listen to the band. Watch for musicians trying to make eye contact with you. During the break between dances, ask the musicians about the monitor settings. Need more piano? Less accordion? If you're a dancer/sound operator, keep a close eye on the band for at least the first three dances.

◄ **Keep your ears on the sound**

The sound of the band can change during an evening, so keep your ears open. You may need to adjust for conditions such as:

- Musicians playing louder as they get warmed up
- Musicians switching to different instruments (you may want to coach them to "back off" or "move in on" with the infrequently-used instrument)

- Larger groups of dancers (more bodies absorb sound and more feet make more ambient noise)
- Heat and humidity (as you get more experienced, you may want to adjust the EQ to compensate as the room gets "warm and wet")

When you're ready to dance

Sound operators for traditional dances are frequently unpaid volunteers. One of the perks of being the sound person at dances, therefore, is being able to join in as a dancer after the sound is set up, checked and running. When that time comes, join the set closest to the sound board as the first No. 2 (inactive) couple. (Tell other dancers that you're the sound person. You might need to make a quick adjustment. Other dancers won't object. This will give you the chance to "jump in" after the first round of the dance and make a change if an adjustment is needed.

Big Band Sound

Doing sound for pick-up bands is a great way to gain experience and earn sincere appreciation from musicians. If you can have things set up and ready to go as they start to arrive, you'll already be a major contributor to the evening's success. Be assertive and build a good gain structure, and you're going to be in good shape for the mix. Here are my suggestions for ways to cope with the special challenges you will have working with big bands.

- **No chairs and no musicians on stage until you have your microphones set up and checked.**

I said it before and I'll say it again: It's much, much easier and more efficient to keep chairs out of the way while you prepare sound for an open band. If you put chairs on the stage, it takes much longer to run cables and do your line checks *(especially when the musicians start arriving and sitting down before you've finished cabling and testing)*.

- **Don't worry about where people with DIs will sit.**

Set up microphone stands and microphones where you expect musicians to be. If someone with a DI sits down, use the microphone cable you have already laid out and tested and strike the microphone setup.

- **Develop a standard setup and modify as needed.**

I do sound for two open bands that typically have as many as thirty musicians in three rows sitting five, six and seven across. But you can never be sure who will show up and who will sit where. So, set up microphones in a standard arrangement, and let the musicians move things around for themselves. (See tip on running microphone cables). Set up microphone stands first, run the cables and then clip on the microphones.

Our standard setup for pick-up bands in the Princeton/Philly area has evolved to place the keyboard in the center of the front row. For the larger, Philadelphia area band we set up three microphones on either side of the keyboard in the front row (seven channels). Our second row of musicians has two microphones on each side of the stage (four channels). We put one microphone on each side for the third row (two channels). When you include the caller, we have fourteen channels laid out. We use the additional two channels on our sixteen-channel mixer as needed, and we can support one or two additional instruments, in a pinch, on the ST1 and ST2 inputs. It can be something of a three-ring circus but it's a challenge and fun.

- **Snake 'em if you got 'em.**

An audio snake can be very useful when working with an open band. It shortens the microphone cable runs and makes it easier to keep cable runs neat and tidy

- **Focus on the basics.**

You need a good rhythm instrument as the foundation for the mix—usually the piano. And you need a strong lead—usually a fiddle. Everything else may sound a little mushy at the start; you'll work it out. If the piano player is weak and you have a strong rhythm guitarist, be sure to get your guitar player on microphone and into the mix. Where you have strong players, highlight them. If a weak player is on a microphone, move the volume to a comfortable level where it blends in.

- **Big band monitor settings.**

Double down on your focus on rhythm. Typically, you want everyone on stage to be able to hear the keyboard and/or rhythm guitar. You can add additional instruments into the monitor mix as you go. As with any monitor mix, communicate with the musicians and find out what they want.

- **Be calm. Be decisive. Be patient.**

Pick-up bands can breed a lot of energy/excitement and requests to the sound person can come fast and furious. Focus on the basics. Stay calm. Prioritize. Do one thing at a time.

- **One musician, one microphone is best.**

I've had times when three and four musicians would crowd around one microphone like an old-timey string band. Some sound people I've worked with as a caller have set up one microphone in front of three or four fiddlers—an "area mic" approach. These would seem like nice ways to get more musicians in the mix, but area microphones create problems. You will end up with microphones set at a very high gain levels in comparison to an individual

Big Band Sound

instrument microphone, and you'll be worried about feedback. All in all, one musician per microphone is best.

◀ **Get strong musicians in the mix.**

What do you do if a strong musician arrives and all the microphone channels are in use? One strategy I pursue, to accommodate a late-arriving musician who can really add a lot to the overall mix, is to "borrow" a channel from someone else in the band. If I've got two guitars in the back row, I'll set up a different microphone and cable, unplug a guitar and put the new musician in that channel. Obviously, you don't want to hurt the feelings of another player. My answer on this point is just not to say anything. If at some point in the dance a player determines that his/her microphone isn't live, it will be identified as a technical difficulty. Usually, an unplugged musician will never be aware of the change. You and the dancers, however, will be aware of another strong player in the mix.

The Incompleat Sound Operator

Putting Your Gear Away

Putting the equipment away properly makes the next setup easier.

WARNING: Mind the instruments.

What's more precious: A musician's instrument or his/her first born? Sometimes, it's a toss-up. ***Whenever you are working on the stage while instruments are present, be very, very careful.*** If possible, do not touch the instruments; let the musicians move them. At the end of a performance you ideally want the musicians to move their equipment and put instruments safely away. But . . . sometimes the performers get distracted. And you need to move things along.

If you can get the musician's attention, ask for help. If the performer isn't available, but the instrument is on a stand, carefully move the stand and instrument out of possible harm's way. No stand? Then move the instrument to a safe place. And be careful.

◂ **Have a process.**

Here is my routine:

- Turn the power off.
- Get microphones first, put them in their storage case and put it aside. You can do this while musicians are putting away instruments.
- Next: Speakers, speaker stands, microphone stands.
- Then: Other gear like powered monitors, DI boxes, and keyboard/piano.
- Then take care of the board. Put on the cover or case. Get it ready to be put away.
- Then, using proper coiling procedures, coil your cables.

◂ **Teach people how to coil cables.**

Always start with the end of the cable where the cable tie is located. This will be the plug end of an AC cable or the "pin" end of a microphone cable. Alternate the direction of the cable loops as you make your coil using the "over-under" technique.

This is where a picture is worth 10,000 words. Look on YouTube for "How to Coil Cables" video by the London School of Sound. A clear explanation and two good techniques.

Note: Some cables—notably AC power cables large and small—come bundled with tight turns in them making them nearly impossible to coil properly. Even if you lay them out, stretch them and try to work out the kinks, they are unmanageable. Here are two techniques for dealing with this:

First and best: Leave them in a hot car on a summer day. They will relax significantly. Stretch them out, coil them gently. Repeat if necessary for a few days.

Method two: Put them in a warm oven (175°F) for two or three minutes. Remove, stretch and repeat. I'd only do this in a pinch and only with small cables like computer and powered speaker power cords. (The hot car method is preferred, but this one works, if done carefully.)

- **A cable without a cable wrap is faulty.**

If a cable or extension cord does not have a cable tie, it should be marked as faulty and put aside for "repair." Keep a roll of hook/loop (like Velcro®) ties and a supply of small electrical cable ties with your sound tools. Put the cable tie on your cable (at the plug end microphone cables and AC power cords). Use the electrical cable tie to anchor the Velcro® tie in place, so it's not sliding around on the cable.

- **Avoid gunk and fight grime.**

Avoid the use of ordinary masking tape and other adhesive tapes that leave a sticky residue. If your cables or other equipment are gunked up with old tape adhesive or other gooey stuff, get a light solvent and clean it up (I use naphtha—Ronson lighter fluid). Keep the board clean of leftover tape goo and sticky stuff, too. It's more fun to work with clean equipment.

Visual Glossary
Quarter-inch Connectors (a/k/a Telephone Connectors, Audio Jack, Headphone Jack)

A family of audio connectors that dates back more than a century. The first quarter-inch connectors were invented for use with telephone switchboards. They were designed with two and three conductors. The 1884 patent for one connector referred to a "jackknife" connector and the receptacle for the connector became known as a jack.

Quarter-inch size connectors are widely used in audio systems today for speaker and instrument connections. Additionally, there are similar 3.5 mm and 2.5 mm connectors. The 3.5 mm connectors are called eighth-inch connectors (though they really are not) and are encountered by sound operators when they are called on to connect consumer gear such as mobile phones, tablets, computers and boom boxes to audio systems. Most quarter-inch connectors used for sound reinforcement systems are monaural. The plug will have two connection points: the "tip and the "sleeve." Consumer audio plugs—the eighth-inch plugs you may encounter—have stereo connections; three contact points: "tip," "ring," and "sleeve." Getting a proper connection between a stereo sound source and an audio board can be tricky. If you think you may need to connect a consumer audio device to your board, it's a good idea to have reliable, tested stereo to mono converters in your gear.

Quarter-inch connector. Note this one is mono; two contact points—the tip and sleeve with one (black) insulation ring.

Eighth-Inch (3.5 mm) connector. Note that it is a stereo connector with three contacts: Tip, ring and sleeve, and has two black insulation rings.

XLR Connectors (a.k.a Cannon Connectors)

The forerunner of the modern XLR connector was invented by James Cannon of the Cannon Electric company, now part of ITT Inc., In the late 1940s. The Cannon X connector did not have a latch to secure it when used as a microphone cable, so a latch was added around 1950 and it was known as an "XL" connector. After the addition of another improvement, embedding, the female connector in an (R for) resilient polychloroprene compound (neoprene) the part designation was updated to XLR. When other companies began producing similar connectors, the XLR designation stuck. Some audio sites say that XLR stands for "External Line Return" but from what I've read, that seems to be an audio urban legend.

XLR connectors are the audio industry standard for microphone lines; providing a secure connection that is shielded from electrical interference that can't be easily disconnected by accident. They are also used for connections between the mixing board and powered speakers.

RCA Connectors (a.k.a Phono Connectors)

RCA Connectors were invented in the 1940s by the Radio Corporation of America for use in in making internal connection in home phonographs. Inexpensive, simple and reliable, they have become commonplace in home audio/video applications. Many audio boards will have RCA input connectors for "record in" and "record out," making it easy to connect audio players and recorders to the board.

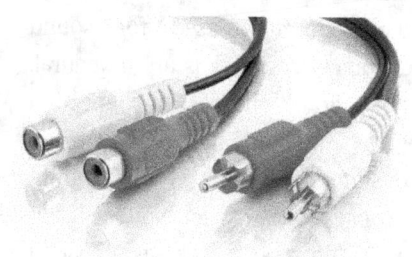

Visual Glossary

speakON® Connectors

speakON® is the trademarked name for twist and latch speaker connectors created by Neutrik AG. They are widely used in pro audio systems for connecting loudspeakers to amplifiers. Other manufacturers make compatible products, sometimes calling them as "speaker twist connectors" or similar names.

Decibel

A decibel is one tenth (deci) of a bel (0.1 bel), a scientific unit named after Alexander Graham Bell. A decibel scale is used to measure sound energy, as well as a number of other power-related forces in the fields of electronics and optics.

The relationship between human hearing and sound energy is complicated. We can hear a vast range of sounds, but our hearing isn't linked directly to the pure physics of sound. While a scientist might measure the doubling of sound 3 dB (though other experts say it is 6 dB), human perception of a doubling of loudness is about 10 dB. Perceived loudness varies based on other factors such as distance and resonance. And to make things more complicated, we hear different sound frequencies differently. We notice increases in mid- and high-frequency sounds more than low frequency tones.

The decibel scale is used on sound boards and in sound meters to create a smoother, more linear way to think about sound. The key takeaway, in my opinion, is that sound is funky. You have to depend on your ears—and ears of the musicians and other people around you. Once you establish comfortable sound levels, a dB meter can help you maintain them consistently. Note the "dB scale" in the diagram on page 42.

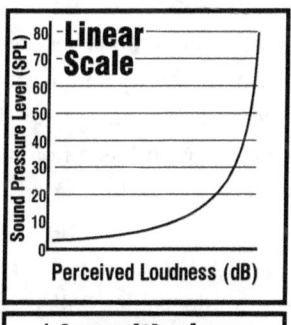

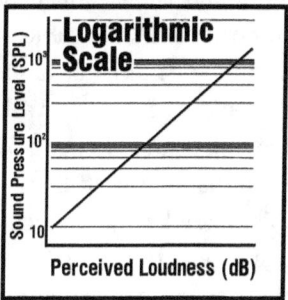

Cable Tester

If you have a problem with a cable (quarter-inch, eighth-inch, XLR, Speakon, RCA and mores) these battery-powered boxes can help you find out where the problem is and help you decide whether to fix it or toss it. And if you decide to fix a cable, they give you a quick and easy way to see if you made the repair correctly. They are available in the $20 to $30 price range.

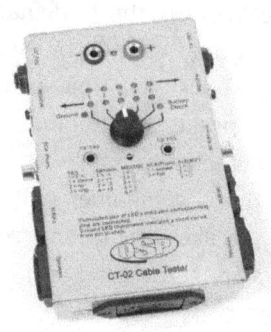

Cable Wrap

A cable without a cable wrap to keep it coiled correctly is defective. The cable wrap shown here is an inexpensive type manufactured by Velcro® that comes in big rolls available at the big home improvement stores like Home Depot. Note that the gray, cable wrap is secured in place with a small, black electrical cable tie. The Velcro® wrap has a hole used to secure it to the audio cable. The electrical tie goes through the hole and around the audio cable. Pull it snug and trim, and your cable wrap won't be sliding around on the audio cable. It will be secured at the plug end of the cable where it belongs.

Bonus tip: Carefully trim the end of the electrical tie flush with the molded ratchet mechanism using a mat knife or razor. If you use wire cutters to trim it you will leave a short, sharp little stub that can cut someone handling the cable.

Visual Glossary 61

Audio Snake

Audio snakes provide a bundle of microphone lines in a single, thick cable. Many snakes also include "returns" that allow you to send signals back from the board to the snake's plugging. Most frequently, returns will be used to provide audio feeds for stage monitors. You only have to roll out one cable from the band to the board. Then you can use short cables to connect from the snake to your microphones and monitors.

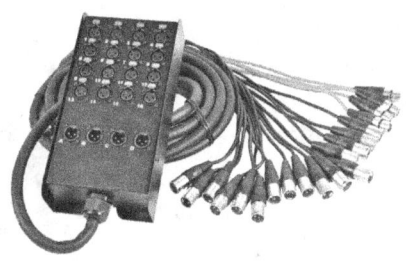

Benefits: Faster set up and simpler, less messy cable runs. And audio snakes are not terribly expensive. A snake might be thought of as "pro" gear but it could be a very beneficial investment for a small group hosting regular performances that require sound support.

Shure SM57 and SM58 Microphones

The SM57 (top photo) is the microphone you will encounter most frequently in acoustic music settings. It is, first of all, a very good quality microphone. It reproduces sound faithfully. Additionally, it is rugged, relatively inexpensive and versatile. With a foam wind screen, it is a very good vocal microphone). All in all, it's a great value.

The SM58 (lower photo) is designed for use as a vocal microphone. It includes a built-in wind screen and delivers consistently good performance for a broad range of voices. The SM58 is the default vocal/caller's microphone for many venues and sound operators.

Personal Monitor

The newest generation of personal monitors provide you with a full sound system in a compact package. Monitors like the one shown here have three or more inputs (channels), very good speakers, and basic HI, MID, LOW equalization controls. They can be used as personal stage monitors for vocalists, a caller's monitor for dances, personal monitors for musicians, or fill speakers for dead spots in a hall.

My current favorite use: A sound operator's monitor next to your board. Put one on a microphone stand (the monitors come with adapters for that) and cable it into the headphone jack on your board. I find it much better than working with headphones for checking pre-fade listen on channels and hearing the monitor mix without leaving the board.

Resources

TradSound E-Mail Discussion List

If you want equipment recommendations or have any sound-related questions and you want a well-informed, unbiased answer, there's no better resource than the TradSound e-mail discussion list hosted at sharedweight.net. Here is a description of the list and its mission:

> *TradSound exists to share information, experiences, techniques about people using intelligence, acoustics and audio equipment to optimally reinforce voices and instruments at community trad music events, including dances, concerts and festivals.*
>
> *Most of our subscribers are audio techs, though we also welcome interested musicians, singers, callers, organizers, dancers and listeners.*

To subscribe to the list visit sharedweight.net/tradsound and follow the "how to join" directions.

All Mixed Up (Book and Website)

Bob Mills, the Vermont-based recording engineer, sound specialist, author, consultant and musician wrote a very helpful book.

All Mixed Up is designed to "teach fledgling sound operators the basics of the craft," and to serve as "a useful resource for experienced sound techs." The book is available online through the Country Dance and Song Society at CDSS.org. You can call CDSS to place an order at 413-203-5467. *All Mixed Up* is also available online at bobmills.org.

Equipment Manufacturers' Manuals

The people who make the gear we use put a lot of time and effort into producing user guides, operating manuals, and other documentation. These documents, almost all available online, provide valuable information. Get the manufacturers' names and model numbers for the gear you use, go online and download the documentation. You'll be surprised by how much you'll learn.

Northern Sound and Light (McKees Rocks, PA)

A great sound equipment supplier. I always call and talk to someone when getting quotes. Great prices and great service, even for very small customers. More info at northernsoundandlight.com

Sound Checklist

Before the Dance/Event
☐ Talk to the band
☐ Get a "picture" of how the band sets up
☐ Have a Sound System Inventory?
☐ Do everything you can to avoid the "last-minute rush."

Set Up
☐ Put chairs in place for musicians
☐ Put the "big stuff" in place (board, speakers, piano)
☐ Keep the board away from an acoustic piano
☐ Accommodate the callers
☐ Set up microphone stands for easy adjustment
☐ Use all the same microphones for musicians if you have enough
☐ Set up an extra microphone stand and microphone
☐ Do NOT Put chairs in place for open bands

Cabling
☐ Start with AC Power, then speakers/monitors, and then microphone cables
☐ Start cable runs at the board
☐ Label your microphone cables
☐ Keep cable runs neat
☐ When you plug mic cables into the board, go left to right
☐ Label the channels
☐ Check it all out
☐ If it's broke: label it and fix it or throw it out
☐ Got tools? Don't be caught without a screwdriver, pliers, battery tester, flashlight or wrench

Working with Musicians
☐ Communicate, communicate, communicate
☐ Even in the dance/folk world, you will run into "divas." Be nice, try to meet their demands, and don't worry about it if they are impossible. Do your best. That's all you can do. And by the way, as a sound person, don't be a diva.
☐ Don't take any negative comments personally

Feel Free to Photocopy these Pages

Setting Microphones for the Band
☐ When time is short, get the directs set first
☐ When time is really short, give musicians microphones and sort out the problems with the directs later
☐ Setting sound levels—gain structure
  1 Set the board to "Neutral"
  2 Set masters at "unity gain"
  3 Set channel slider at "unity" and "get in the ballpark" with channel gain control. Repeat process for each channel
☐ Make final adjustments with channel controls and monitor controls—don't change channel "preamp" gain control unless absolutely necessary
☐ Fine tune monitors for the musicians
☐ Fine tune house sound

Equalization
☐ Less is more
☐ Experiment with caller sound during walkthroughs

During the Dance
☐ Walk the hall
☐ Pay attention to the musicians
☐ Use a sound meter if you have one
☐ Ask for comments—be open to feedback from others

Big Bands
☐ No chairs or musicians on stage until sound is ready
☐ Focus on the basics
☐ Be calm
☐ Get strong musicians in the mix

Packing Up After the Dance/Event
☐ Have a process
☐ Be extra cautious around musical instruments
☐ Teach people how to coil cables correctly
☐ Fight gunk and grime: If gear gets dirty clean it up right away

Feel Free to Photocopy these Pages

Feedback (the Good Kind)

It is my hope that this volume will be something of a collaborative effort; that it will stimulate thoughtful comments, corrections, additions and other recommendations for improvements from you, dear reader.

Modern book production technology, especially print-on-demand publishing, facilitates revised and updated editions. I hope that this first iteration of *The Incompleat Sound Operator* will, thanks to contributions from you, be quickly replaced by a second edition.

All feedback along these lines will be greatly appreciated. Please send thoughts to me at hedgehoghousebooks@gmail.com. If it's a complex recommendation, just give me the gist and include your contact information so I can reach out to you for a full explanation. I would truly like to make this book the best that it can be.

I also invite you to join the Hedgehog House mailing list for occasional messages about this and similar publications. Visit the Contact page at HedgehogHouseBooks.com to sign up.

Sincerely,

Ridge

www.ingramcontent.com/pod-product-compliance
Lightning Source LLC
Chambersburg PA
CBHW052205110526
44591CB00012B/2093